I0483194

ISBN – 13 978-1537767727
ISBN – 10 1537767720

Printed in the United States of America

THE
INTELLIGENT

SYSTEMS

Prof. Dr. Iwasan Kejawa
Miami Dade College

@@@@@@@@@@@@@@@@@@@@@@@@@@@@@@
@@

The scientific educational aspects of mathematics are explored in this book. It describes how to formulate ideas and functions mathematically through critical thinking and brainstorming. Creating mathematical formulas can be derived through thinking of a problem or situation. We can create formulas by applying theoretical, technical, and applied knowledge. The knowledge derives from brainstorming and actual experience can be represented by formulas.

It is intended that the book be used by an audience of average knowledge of solving mathematics and scientific education intricacies

Author

Dr. Iwasan Kejawa born in Erekiti, Okitipupa, Nigeria, Africa.

He is an educator and writer by profession. He has earned Bachelor of Business Administration in Computer Systems, Master of Science Computer Sciences and a Doctor of Education Degrees.

Dr. Kejawa is the author of the books, Education: Leadership in Positive Ways; Strategic Analysis of Education Fundamental for Modern Society; Raw and Pure Education; The Turbulent Economy: Achieving Success in Tumultuous Education; Reaching the Heights. Computers in Society: the world of science and Technology. Mathematical Intelligence: the art of critical thinking and Fundamentals of Education.

He has partaken in various researches both corporately and academically for number of years. He is a professor of Computer Sciences at Miami Dade College and Broward College

Mathmatetical Expert Systems Analysis and Education

Mathmatetical Expert Systems Analysis and Education

Mathmatetical Expert Systems Analysis and Education

Table of Contents

CHAPTER TWO

EQUATIONS/FUNCTIONS

SIMPLE

SIMULTANIOUS

QUADRATION

POLYNOMIALS

MONOMIALS

BINOMIAL

TRINOMIAL

CHAPTER THREE

THEORETICAL APPROACHES

ELIMINATION

SUBSTITUTION

INTEGRATION

CHAPTER FOUR

APPLIACTIONS

BUSINESS

SCIENTIFIC

TECHNICAL

SUMMARY

FOREWARD

Creating mathematical systems formulas can be derived through thinking of a problem or situation. We can create formulas by applying theoretical, technical, and applied knowledge: the knowledge derives from brainstorming and actual experience that can be represented by formulas.

It is intended that this book be used by an audience of average knowledge of solving mathematics and scientific intricacies. It can be used in secondary schools and colleges at the introductory level of simple,

Mathmatetical Expert Systems Analysis and Education

at times complex problems in a Mathematics course and computability and solvability in a Computer Science course.

In this book, various ways of creating formulas and solving simple, at times complex problems are explored. Scenarios are introduced and answers are provided. Reading this book thoroughly and getting familiar with the logic behind each situation of a critical problem will enable the readers to formulate techniques of solving problems by creating or representing mathematical formulas. The book also explained how the formulas presented can be solved to attain end results.

Acknowledgment

My gratitude to all contributors to this book directly and indirectly. Thanks to the entire faculty of the Department of Mathematics and Statistics at the Bernard M. Baruch College, the Department of Computer sciences of the School of Engineering at the City College of New York of the City University of New York. I also applaud the faculty members that guided me through my post graduate courses in Management at the New York Institute of Technology.

I would like to express my thanks and sincerity to the staff and faculty of the Fischer School of Education and Human Services of the Nova Southeastern University for the knowledge I acquired during my affiliations with the institution as a Doctoral Student

As it is often said, "A friend in need is a friend indeed", vice versa. Special thanks and gratitude go to all my friends, family, and relatives.

Mathmatetical Expert Systems Analysis and Education

Written by Iwasan D. Kejawa

THANKS TO VIORIS AND JUDITH

- 12 -
Mathmatetical Expert Systems Analysis and Education

Mathmatetical Expert Systems Analysis and Education

INTRODUCTION

Mathmatetical Expert Systems Analysis and Education

To create system formulas one must represent the objects and the functions with variables. Formulas are used to analyze various businesses, science and engineering problems as well as other problems encounter in our daily lives. We often create formula by our actions which we are not aware off. Our activities or actions can be represented systematically by formulas. We do apply formulas in everything that we do. The systems by which we do things are formulas: Formula can be categorized as written plan to solve a problem or to accomplish a function. It is usually in Mathematical or Algorithmic form.

FOR INSTANCE:

If 16 items are to be sold for 5 dollars each and 3 of the items are sold for 7 dollars each. What is the total price of all the items? How many items remain if only 10 of the items are sold?

Representations:

Let   A = the number of item1

        B = the number of item2, etc.

Mathmatetical Expert Systems Analysis and Education

Z = Remainder

S = number of items sold

Then Formulas:

T = AP or A1 x P1

Where A = number of each item1

And B = Price of each item1

Z = A – S

A = OBJECTS, P and S are functions. And

P (A) = T, P(S) = Z

T and Z are outcomes or end results.

Solving for T and Z:

A = 16

P1 is the first instance which is equal 5

P (1) = 5, P (2) = 7

P2 is the second instance which is 7

S = P (1) + P (2) = 12

S= 12

Mathmatetical Expert Systems Analysis and Education

T1 is the total of first instance and

T2 is the total of second instance

$$T1 = (16 - 3) \times 5 = \$65$$

$$T2 = 3 \times 7 = \$21$$

$$T = 65 + 21 = \$86$$

$$Z = A - S$$

$$= 16 - 10$$

$$= 6$$

PERFORMANCE EVALUATION FORMULAS

To create and calculate performance we will multiply activity by action.

Performance evaluation can also be regarded as combinations of action and object

Formula Representation:

Mathmatetical Expert Systems Analysis and Education

Let p = Performance, Action = A and Activity = B

Then Formula:

$$P = A + B$$

For instance:

Imagine an automobile that travels at a constant speed of 80 miles per hour and it took 5 hours for the driver to get to his or her designated destination. What is the over all performance of the automobile.

Since A= action and B activity

Activity = 80

Action = 5

Then

Overall Performance = 80 x   5

= 400 miles per 5 hours

Mathmatetical Expert Systems Analysis and Education

Or to put another way,

We say:

Rate x Time = Distance

80mph x 5hr = 400

Mathmatetical Expert Systems Analysis and Education

CHAPTER ONE

FUNCTIONS

ADDITION FUNCTION

Addition is the combination of two or more entities.

Formula Representation:

$$A + B + \ldots\ldots n$$

Where n varies

For instance

Let assume we want to add six numbers OR more.

Then Formula:

Function (F1) = n1 + n2 + n3 + n4 + n5 + n6 + ... N

(F2) = (n1, n2, n3, n4, n5, n6, n

Example:

Subtraction Function

Mathmatetical Expert Systems Analysis and Education

Subtraction is the elimination of one or more variables

from an entity

FORMULA REPRESENTATION:

$$A - b - \ldots N$$

FOR INSTANCE

We can eliminate or take away

several entities or items from a Whole.

Thus Formula:

$$\text{Function (F)} = N - n1 - n2 - n3 - \ldots n$$

Example: To subtract the number 23 and 25 from whole

number 100.

$$\text{Function (F)} = 100 - 23 - 25$$

$$= 52$$

MULTIPLICATION FUNCTION

Mathmatetical Expert Systems Analysis and Education

Multiplication is the addition of a variable n number of times.

For instance

The addition of Y n number of times is

Y (n)

FORMULA REPRESENTATION:

Then Formulas:

Function (F) = Y x n

Y * n

Y. n

= Y1 + Y2 + Y3 + ... + Yn

Example: To add a number 1000, 1250, 515 and 6000 series of time

Function (F) = 1000 + 1250 + 513 + 6000 = 8763

DIVISION FUNCTION

Mathmatetical Expert Systems Analysis and Education

The division function is the partition of a whole (Y) of

entity n into x number of times. It is the number of times

n occurs in Y or the occurrence of n at x number of

times.

FORMULA REPRESENTATION:

FUNCTION (F1) = Y / n

Thus Formula:

(F1) = n (1), n (2), n (3), n(x)

(F1) = x

Example:

To divide the number 12 by 2 will be

F = 12 / 2

F = 2(1), 2(2), 2(3), 2(4), 2(5), 2(6)

F = 6

QUOTIENT FUNCTION

Mathmatetical Expert Systems Analysis and Education

Quotient function is how many times an entity varies

after the entity is divided by its divisor,

For Instance

Let assume N is divided by y at z times

FORMULA REPRESENTATION:

Then Formulas:

FUNCTION (F) = N / y (1). N / y (z)

Where Y (z) not equal 0

Thus:

y = divisor of

N such that y1 x y2.... is = N

z =

occurrence of y

FUNCTION (F) = Z

Example:

To find the quotient of a number 8

The divisors 1, 2, 4 and 8

Mathmatetical Expert Systems Analysis and Education

Since   8/4 =2 and 2 x 2 x 2 = 8

Using the above that was given earlier, then:

Function (F) = 8/4(1) X 8/4(2) X 8/4(3)

Quotient = 3

AVERAGE OR MEAN FUNCTION

An average function is the sum of a group of numbers divided by the number of times the number existed in the sum. The average function of six numbers represented as A, B, C, and D, E, F would be for example being formulated as follows:

FORMULA REPRESENTATION:

Formula:

Average Function (F) = (A1 + B2 + C3 + D4 + E5 + FN) / n

Mathmatetical Expert Systems Analysis and Education

A, B, C, D, E, F = set of numbers

n = number of occurrence

Example: A

An average function of a group of numbers 2, 4, 1, 3, 2

First count the number occurrence of the number group:

$y(1) = 2$, $y(2) = 4$, $y(3) = 1$, $y(4) = 3$, $y(5) = 2$

n = 5

Thus average Function

$(F) = (2(1) + 4(2) + 1(3) + 3(4) + 2(5)) / 5 = 12/5$

$(F) = (2 + 4 + 1 + 3 + 2) / 5$

$= 12 / 5$

Average Function (F) = 2.4

Mathmatetical Expert Systems Analysis and Education

ROOT FUNCTION

Root function is number of times a divisor of giving

number is multiply by itself to equal to the whole

number.

Representation Formula:

$F = N/n(y) \times N/n(y)$

N = the giving number

n = the divisor,

y = number of times it occurs

Example A:

To find the root function of the number 16

First find all the possible divisors, then

How many times it is multiply by itself to arrive at the

same number?

Function (F) = 16/4(1) x 16/4(2)

= 4 x 4

Mathmatetical Expert Systems Analysis and Education

$$= 16$$

$$y = 2,$$

$$n = 4,$$

Therefore, we say that 4 is the square root 16

Since when it is multiply by itself yield 16

Alternatively:

Function (F=) 16/8(1) x 16/8(2) x 16/8(3) x 16/8(4)

$$= \quad 2 \quad x \; 2 \quad x \; 2 \quad x \; 2$$

$$= 16$$

$$y = 4$$

$$n = 2$$

We arrive at the same solution except that y and n are

Inter-changed. We may say that 2 is the fourth root of 16

Example B:

If we are asked to find root function of 27?

N = 9, the divisor = y,

Number of occurrence = n

Mathmatetical Expert Systems Analysis and Education

Then the formula =

   N/y (1). N/y (2). N/y (3) and so on .... (n)

Therefore,

   = 9/3(1). 9/3(2). 9/3(3)

   3.   3.   3

   = 27

   y = 3

   n = 3

Therefore, the root is equal to: 3

Here we say 3 is the cube root of 27

Example C: Find the root number of 36?

   Solution: N = 36, possible divisors is 9, 3, 4, 6

   Therefore, y = 9, 3, 4, 6

When y = 9

Then using the formula, it is 36/9(1). 36/9(2). 36/9(3)

Which equals?

Mathmatetical Expert Systems Analysis and Education

y = 9

= 4. 4. 4

= 64

Therefore 4 is not the root of 36 because 64 is large that the actual number. (36)

When y = 3

We have 36/3(1). 36/3(2) 36/3(3)

= 12. 12. 12

= 1728

Since 1728 is larger than 36

Then 9 is not the root of 36 because 729 is larger than 3

When y = 6

We have:

Mathmatetical Expert Systems Analysis and Education

= 36/6(1). 36/6(2). 36/6(3)

= 3. 3 . 3

= 27

y = 3

n = 6

Since 27 is less than 36 then 4 is not the root

of 36

When y = 4

We have:

= 36/4(1). 36/4(2)

= 6. 6

= 36

Since this is equal to actual number then 6 is the

root of 36, we may say that 6 is the square root of 36

because when 6 is multiply itself we get the whole

number 36

Mathmatetical Expert Systems Analysis and Education

CHAPTER TWO

EQUATIONS

Equations can be categorized as (1) Simple Equation, (2)

Simultaneous Equation (3) Quadratic Equation, (4)

Monomial Equation (5) Binomial Equation and (6)

Polynomial Equation. These equations can be regarded

also as functions:

- **Simple Function**

- **Simultaneous Function**

- **Quadratic Function**

- **Monomial Function**

- **Binomial Function**

- **Trinomial Function**

- **Polynomial Function**

Simple Equations

These types of equations can be created by multiplying,

dividing, subtracting or adding fixed numbers and

variables. A simple equation function may be in the

following terms, a fixed number n and variables y an x:

Mathmatetical Expert Systems Analysis and Education

Representations:

$$\text{i.} \quad F = ny$$

$$\text{ii.} \quad F = n + y$$

$$\text{iii.} \quad F = n/(n + y)$$

$$\text{iv.} \quad F = n\text{-}y$$

$$\text{v.} \quad F = n - y - x$$

The above examples are just possibilities of simple equation functions. There are a lot of simple equation functions.

Example: A

Let a given variable number be y and a fixed number be 8 and a function be equal 16. Now we can derive the appropriate possible simple equations.

$$n = 8$$

y=variable number

$$F = 16$$

The equations for the above example may be as follows:

1) $16 = 8y$

or

2) $8y = 16$

Mathmatetical Expert Systems Analysis and Education

3) $8y/8 = 16 / 8$

4) $y = 2$

5) $16 = 8 + y$ or $y + 8 = 16$

6) $y = 16 - 8$

7) $y = 8$

8) $16 = 8/(8+y)$

$16(8 + y) = 8$

$128 + y = 8$

$Y = -120$

9) $16 = 8 + y$

10) $-y = 16 - 8$

11) $-y = 8$

12) $y = -8$

With example A, let another fixed number (x) to be 3, then

13) $16 = 8 - y - 3$

14) $16 - 8 + 3 = -y$

15) $8 + 3 = -y$

16) $11 = -y$

17) $y = -11$

Example B:

Let a given variable be equal to 45 and a fixed number equal 65 with given function 3. What are possible simple equations?

$$T = 65$$

$$r = 45$$

$$z = 3$$

The possible number of equations for the above example can be enumerated as follows:

1). $45r + 3z = 65$

2). $3z = 65 - 45r$

3). $T = \dfrac{65 - 45r}{z}$

Simultaneous Equations

These types of equations are formed with two different or same types of functions. Each function can be in form of two equations. Both of the equations may also

Mathmatetical Expert Systems Analysis and Education

be in form of additions, subtractions, multiplications

and divisions.

Representations:

i. $x + y = C$

$x - y = Z$

ii. $x-y = C$

$x- y = Z$

iii. $Ax-Ey = C$

$Ax+By = Z$

iv. $Ax - Ey = C$

$Ax + Ey = Z$

v. $Ax/By= C$

$Ax/By= z$

vi. $Ax = C$

$Ey = Z$

Mathmatetical Expert Systems Analysis and Education

Example A:

If in a process, F1 = 10 and if a given number 4 is multiply by a variable x and added to a fixed number 2 which is multiply by a variable y. In another process F2 = 8 and a given fixed number 2 is multiply by a variable x and subtracted from a given fixed number 5 which is multiply by a variable y. Formulate the appropriate simultaneous equations for this scenario and solve for x and y

Representations:

F1 = 4x + 2y = 10 i

F2 = 2x + 5y = 8 ii

Solving:

In other to arrive at a solution we can multiply equation ii by (- 2) as follows:

Mathmatetical Expert Systems Analysis and Education

$4x + 2y = 10$.........i

$(-2(2x)) + (-2(5y)) = (-2(8))$ which will be equal to:

$4x + (-10y) = -6$ OR $-4x - 8y = -6$

Both equations can be represented as follows:

$4x + 2y = 10$.................i

$-4x - 10y = -6$....................ii

We can now eliminate x and solve for y by adding equation is from equation ii.

Quadratic Equations

These kinds of equations can be termed as functions where a variable is multiply by itself once or more number of times in the equation (i.e varies once or more times) and combined with a constant number or fixed number (Refer to the section on Root Functions on Chapter One).

Mathmatetical Expert Systems Analysis and Education

Representations:

$$F1 = Z(1) \times Z \text{ (n times)} + N = 0$$

$$F2 = Z.Z \text{ (n) times} + N = 0$$

$$F3 = Z' + N \text{ where' equals n times}$$

$$F4 = Z * Z \text{ (n) times} + N = 0$$

Z, n are variables While N is constant.

Example A

Let us look at a function that has a variable number of 6 which varies two times and added to a Constant number 3. What is the solution for the Function and its quadratic equations formulas?

Solving

$$\text{Let } Z = 6$$

$$n = 2$$

$$N = 3$$

Therefore

$$F = Z(1) \times Z(2) + 3 = 0$$

$$= 6 \times 6 + 3 = 0$$

$$F = 15$$

Or

$$F = Z.Z + 3 = 0$$

$$= 6.\ 6 + 3 = 0$$

$$F = 15$$

Or

$$F = Z' + 3 = 0 \text{ where'} = 2$$

$$6' + 3 = 15$$

Or $\qquad 6 * 6 + 3 = 15$

Polynomial Equations

The Binomial Equations are equations where a variable is multiply by self n number of times (that is (i.e.) n times) combined with a variable and constant or fixed number

Representations:

$$F1 = Z\ (n)\ x\ \ Z\ (n) + T + N$$

$$F2 = Z\ (1).\ Z\ (n) + T + N$$

$$F3 = Z' + T + \ N$$

$$F4 = Z *\ Z + \ T\ + \ \ N$$

Z, n, T and 'are variables While N is constant or fixed number.

Mathmatetical Expert Systems Analysis and Education

Example A:

If we have a variable number 10 that varies 2 times

combined with another variable number 5 with an

addition or subtraction of a constant or fixed number 20;

what are the possible formula representations of this

problem and answer to this problem?

Solving:

$$F = Z(1) \times Z(2) + T + N$$

$$Z = 10, T = 5, N = 20$$

$$= 10 \times 10 + 5 + 20$$

$$= 125$$

Or

$$F = Z * Z + T + N$$

$$= 10 * 10 + 5 + 20$$

$$= 125$$

Or

$$F = Z' + T + N$$

Where ' = 2

$$= 10' + 5 + 20$$

$$125$$

Or $\quad F = Z. Z + T + N$

$$= 10. 10 + 5 + 20$$

= 125

ALGEBRAIC EXPRESSION:

An algebraic expression is a combination of ordinary numbers and letters which represent numbers.

Examples:

2) 3(a**2)(b**7)

3) (3x*y + 4*z) / (2*a – c**2)

Monomial Equation

A monomial equation consists of products and quotients of ordinary numbers and letters which represent numbers. It consists of one term.

A term is integral and rational.

Examples are:

6(x**3)y

5x / (3x**2)

-4x**3

Binomial Equation

A binomial equation is an algebraic expression that consists of two terms.

Example:

6x**2 + 7 * x * y

Trinomial Equation

A trinomial equation is an algebraic expression consisting of three terms.

Representations:

    1) $F = Z(1) * Z(2) * Z(3)_{+T-N}$

    2) $F = Z**3 + T - N$

    3) $F = Z.Z.Z + T - N$

Examples:

    $3x**3 + 6x*y - 2$

    $3x.x.x + 6x.\ y - 2$

If x=2 and y=10

Then in

  $3x**3 + 6x*y - 2 = 3(2) **3 + 6(2) * 10 - 2$

        $= 3((2.2.2)) + 6(2.10) - 2$

        $= (3.\ 8) + (6.\ 20) - 2$

        $= 24 + 120 - 2$

        $= 144 - 2$

        $= 142$

Mathmatetical Expert Systems Analysis and Education

COMPUTATIONS WITH POLYNOMIALS

Addition:

The sum is obtained by combining rational numbers

in front of the like terms.

For example: 3X **2 * 2Y ** 3 - 4x **3 * y ** 5

X= 2 + 3 while y = -3 + 5

Subtraction:

The subtraction is done by deduction of rational

numbers in front of the like terms:

Example: 3X **2 * 2Y **3 – 4x **3 * y **5

Would be

X = 2 - 3 while y = -3 + 5

Mathmatetical Expert Systems Analysis and Education

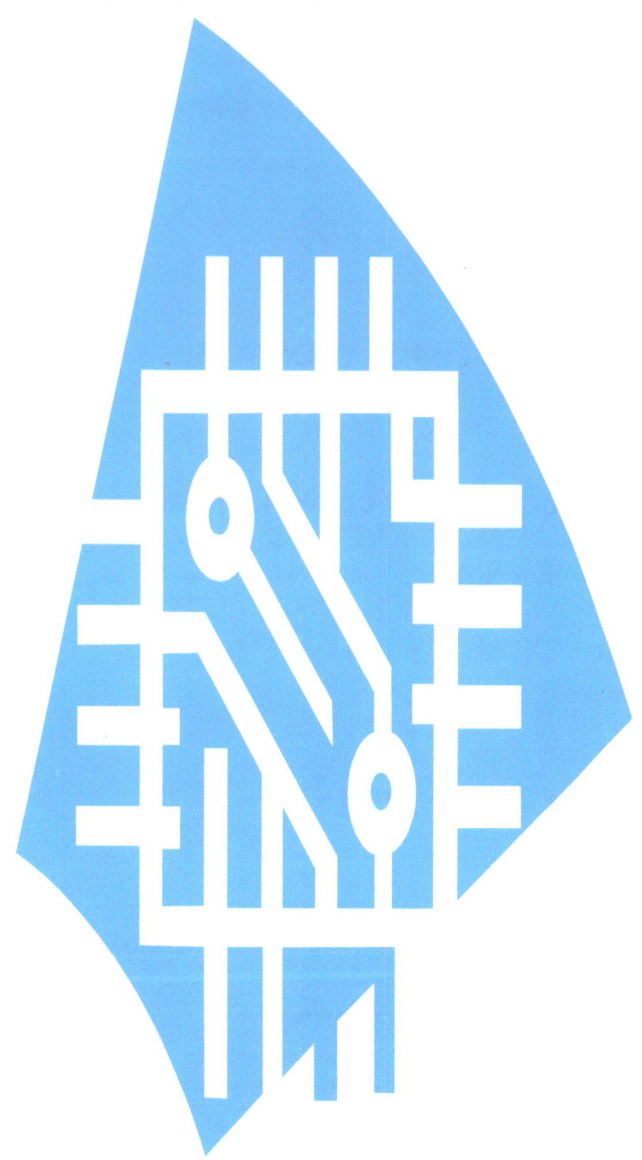

Chapter Three

APPROACHES

The approaches of solving problems or predicaments
are what mathematics functions convey to the world of
science. It is imperative that hypothesis be analyzed,
concluded and integrated. The problem of general
science is that sudden approach is realized with
substances with sudden beliefs.

Eliminate Method

$4x + 2y = 10$ i

$0 - 8y = -6$................ iii

$y = -6 = -6/-8 = -3/4$

$y = -.75$

Substitution Method from equation i:

In order to find the value for x

$4x + 2y = 10$...............i

$4x + 2(.75) = 10$

Mathmatetical Expert Systems Analysis and Education

$$4 \text{ xs} + 1.5 = 10$$

$$4x = 10 - 1.5$$

$$x = 8.5 / 4$$

$$= 2.125$$

Integration Method

F1: $4x + 2y = 10$...............i

F2: $2x + 5y = 8$ ii

Solving:

F1: = $4(2.125) + 2(.75)$

$$= 8.5 + 1.5$$

$$= 10$$

F2: = $2(2.125) + 5(.75)$

Mathmatetical Expert Systems Analysis and Education

= 4.25 + 3.75

= 8

Mathmatetical Expert Systems Analysis and Education

Chapter Four

APPLICATIONS

The applications of formulas are the willingness to

control the situations around the globe. In order to

sharply increase example-nary proprietaries using

formulas we must know when, where and how to applied

formula. The innovative process of mathematical

Mathmatetical Expert Systems Analysis and Education

formulas would have an impact on the behavioral of the

populace. One of the technicalities of eradicating the

impact of innovation is to limit and not to invigorate

exposure of the formula. The applications of

mathematical formulas may have to involve acquiring

necessary skills, experience and practice. The

provisions of simple and positive formula will certainly

improvise the business and scientific applications as

well as technical applications of Mathematics   With

Formulas we can easily solve problems precisely.

Formulas are often applied to the derivations of

solutions which are not made aware to everyone.

Mathmatetical Expert Systems Analysis and Education

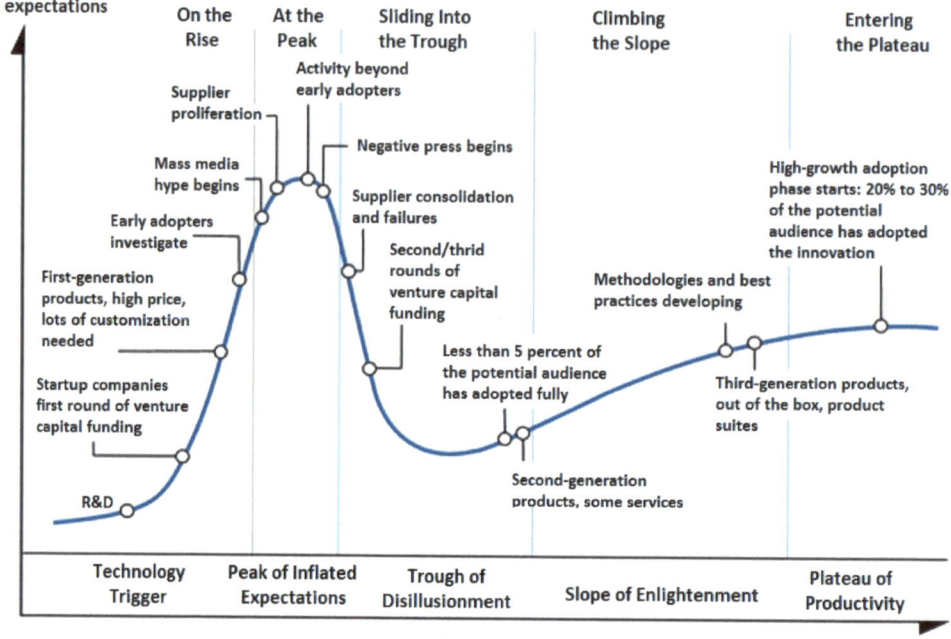

Business Applications

We all used certain formula to carry out activities or to solve problems. Although many businesses of today use computers to solve business problems, we still have to formulate how to interpret the business solutions from these computers. Managers and business computer applications employees do formulate mathematical algorithms that are used to resolve problems manually. In the banking industries for example, mathematical algorithms are used to calculate or process customers' transactions, and in the

Mathmatetical Expert Systems Analysis and Education

insurance companies, formulas are used for projections

and annuity calculations.

Many large businesses depend on the use of

computers to survive in the world today. The benefits of

using computers to solve problems outweigh their risks.

To solve problem, we must first of formulate the

problem by creating algorithms or mathematical formula

and then use our intuitions to manipulate, solve or

process the data or the information relating to the

problem. Today instead of using our intuitions, most of

the businesses use computers to solve problems in all

their endeavors: Accounting, Production Management,

Finance, Human Resources and Operations

Management and so on are some of the areas

computers are applied by businesses.

Scientific Applications

Mathematical formulas are used by scientists for many scientific functions. They are used for predictions and solving problems of physical and unphysical entities or scenarios.

They can be used for example to derive the amount of intake of oxygen in species in a period of time. Scientists applied formulas in the same manner as a business man and technician or engineer. The creations of scientific formulas have made life easier to mankind. The use of formulas to solving scientist problems has made the world a better place to explore various ways of doing things and surviving on earth. Scientific applications of formulas actually lead to progression of human existence, and the physical being itself.

For example, formula can be used to measure the amount of granules a plant would need to grow normally and how many times the plant supposed to be fed.

Mathmatetical Expert Systems Analysis and Education

Technical Applications

Mathematical formulas serve various purposes in the technical environment. Engineers just as scientists and business persons use formulas to carry out or perform their functions. Technical applications of formula enable engineers to design physical entities and implement most functions

The use of formulas is very common in engineering. For example, in the designs of bridges, automobiles, robots, houses and so on, just to mention

Mathmatetical Expert Systems Analysis and Education

a few. An example of the usage of formula by engineer

in building a bridge is to find the capacity and the length

of the proposed bridge by measurement. In

manufacturing automobile for example: the engineers

may use formula to predict how fast an automobile can

travel and the amount of gas it will consume or use.

SUMMARY

PROBLEM SOLVING AS APROACHED TO EXPERT SYSTEMS AND ARTIFICIAL INTELLIGENCE

Problem solving can be attributed to intelligent systems

as well to expert systems. The systems generally play a role

in behaviorism and in artifacts. The perpetuation of objects

may lead to its resilience and its functionalities. The capacity

Mathmatetical Expert Systems Analysis and Education

to which a subject is represented depends solely on the functionalities of the issues. It has come to a place where artifacts are considered as live embodiments. Living objects and non living objects are subjected to rigorous understanding of their environment. For example, a street light knows when to turn on and turn off when it is dark or when there is a bright environment respectively. An object is subjected to scrutiny when the justifications are beyond control of behold. A form of participative objectivity is derived from subjective activities of the living objects.

The knowledge instilled in the livings can be represented as artifacts. So also can non living entities be represented as living entities, thus our world existed in learning and conveying knowledge with the aid of the living and non-living, this justification will make us believed that artificial intelligence and experts systems are both interchangeable. The complexity of acquiring knowledge is based on the sophistication of learning endeavor.

The situations of understanding lively artifacts and the process of creating physical paradigms remain a soul entity in the ages of expert systems and artificial intelligence. Approach of these subjects is subjected to the soul of soul of the era of Plato and Socrates. The breath of the mind is the

Mathmatetical Expert Systems Analysis and Education

elongation of physiological, psychological, sociological

parabolic entities of our times. Those entities with positives

and negatives thoughts can be transferred to non-livings and

livings. But these means there are sensitivities between non

living and the living.

This may lead us to an example of moveable objects in

the world of today. Moving objects depend solemnly on the

view and situation of the beyond. Transportation model, one

example of serving the purpose of experts systems and

artificial intelligence is dated back to the early era or years of

vulnerability and adaptability which is religiously expressed

in various religion peripherals, memorabilia and paraphilania.

An instructional modality may serve to accomplish a goal,

and eliminate instances of confrontation between two

subjects. A transportation problem analysis as oppose to

agricultural and household analysis may result in the

generality of instances where physical entities need to be

moved between a point to another point without any

disruptions, discrepancies or problems.

Mathmatetical Expert Systems Analysis and Education

Below is a Hierarchical Chart for a Transportation

Systems:

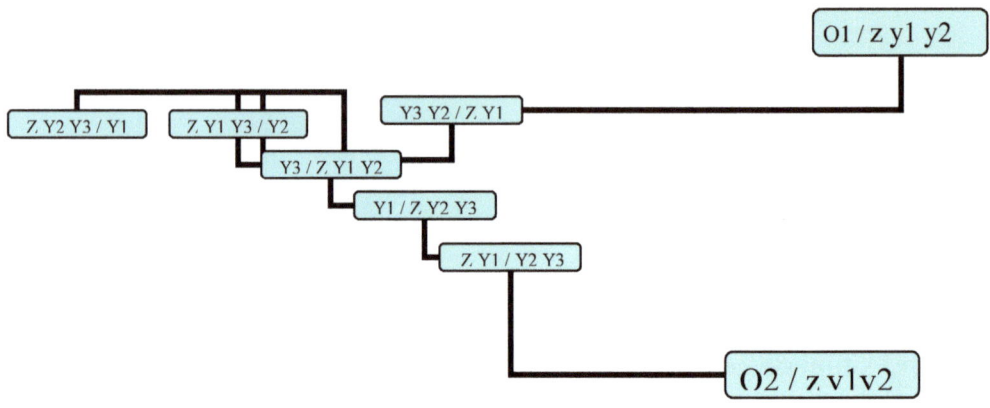

The Followings are Mathematical Formula

Representations for a procedure of

Transportation Systems:

O1 = Departure
O2 = DESTINATION
/ = Boundary between the two points
Z = Sailor or Driver or Pilot or Farmer or Locksmith
(INDIVIDUAL)
Y1 = Object 1
Y2 = Object 2
Y3 = Object 3

Mathmatetical Expert Systems Analysis and Education

(Object 1 is deadly to object 2, object 3)

BEGINNING:
O2 = 0 WHEN O1 = Z + Y1 +Y2 + Y3

O1 = Y3 + Y2 WHEN O2 = Z + Y1
O2 = Y1 WHEN O1 = Z + Y2 + Y3

O1 = Y3 WHEN O2 = Z + Y1 +Y2, O1 = Z + Y3 + Y1 WHEN
O2 = Y2
O2 = Z + Y2 + Y3 WHEN O1 = Y1, O2 = Y3 + Y2 WHEN O1
=Z +Y1

ENDING:
O1 = 0 when O2 = Z + Y1 + Y2 + Y3

There are 16 possibilities of situations and 8 movements of objects and 6 occurrences of presences of the individual (Z) in this scenario.

The surveillance of destructive elements needs to be

addressed to obtain a goal or to accomplish a mission.

An instance when a fisherman has a boat, a rabbit, a

dog and a cat and need to transport them across a river

without any fatal incident is an abominant analysis of the

intelligent systems. In this scenario the fisherman cannot

carry more than one thing in the boat at a time. And if the

rabbit is to be left alone with the cat, there will be an

altercation between the two that may lead to the dead of the

rabbit. This scenario needs to be examined to determine the

ways to solve a problem. The mathematical analysis of the

transportation problem described previously in this text is an

example of the solution to the fisherman problem.

Mathmatetical Expert Systems Analysis and Education

The solutions to the fisherman problem can be discussed in a layman approach. First the fisherman needs to carry the rabbit across the river and leave the cat and the dog alone. Next he needs to leave the rabbit alone at his destination and comeback to carry the cat. He then needs to leave the cat at his destination and carry back along the rabbit.

Next thing for him to do is to carry the dog across the river and leave the rabbit alone at the source location. After he gets to his destination, he is then to leave the dog with cat and comeback to carry the rabbit. This will complete his mission without any problem between or with the species. The problem of the fisherman can also be applied to farmers; where a farmer possesses a grain, bird and a rabbit. So can the problem be applied to a locksmith with a house, a dog, solar, and mental or iron? The solar may be detertrimental harmful to the dog and the mental or iron. If the locksmith is to leave the house for long period of time or to try to relocate to far away town or country and must take with him or her no more than one item and he or she cannot leave the dog with someone or with a veterinarian and neither can he or she have somebody to keep the house in

Mathmatetical Expert Systems Analysis and Education

the house; then he or she will have to follow the same process as the fisherman and the farmer.

For example, if it happens that there is a Toledo or the weather gets too hot during his or her absence, the solar system may cause the mental to be too hot and thereby causing the temperature in the house to be too hot which may cause the dog to be uneasy and result in sickness or in death.

Both the locksmith and farmer problem can be resolved by following the same solution just like that of a fisherman. For the farmer with a goose, grain and cat; if the farmer leaves the goose with the grain, the goose will eat the grain and if he leaves the cat with the goose there will be problems or misunderstandings between the two which may result in the death of the goose.

To solve this problem, the famer will have to take with him the goose, leave the goose at his destination and leave the cat and the grain at the initial departure, then comeback and carry the grain, leave the grain at his destination and carry back the goose to his point of initial departure and leave goose there. He will then have to carry the cat and leave the cat in his destination with the grain and return to carry the goose.

Mathmatetical Expert Systems Analysis and Education

NOW TRY TO:

USE THE ABOVE SIMILAR SCENARIOS TO SOLVE

THE LOCKSMITH PROBLEM???

?

?

?

?

- 65 -

Mathmatetical Expert Systems Analysis and Education

Mathmatetical Expert Systems Analysis and Education

COMPUTERS AND ARTIFICIAL INTELLIGENCE

Artificial intelligence is the representation of human

knowledge by machines or objects. Objects are devised

through expert system to perform various human functions

Mathmatetical Expert Systems Analysis and Education

or sophisticated tasks that cannot be performed or very

difficult to be performed by humans.

Computers may be referred to as artificial intelligence or

expert systems.

The various usages of **Artificial Intelligence and**

Computers can be enumerated as follows:

     a) Business

     b) Engineering

     c) Manufacturing

     d) Farming

     e) Mining

     f) Schools

     g) Hospitals

     h) Households.

Mathmatetical Expert Systems Analysis and Education

1) **The Representation of commonsense Knowledge.**

   (ROBOTS)

    a) **Automobile manufacturing (**Such as in

       assembling and driving).

Mathmatetical Expert Systems Analysis and Education

b) **Operation services (**Such as in household/office chores, Errands deliveries.)

2) **Language Understanding.** (ROBOTS AND COMPUTERS)

a) **Interpretation of simple questions and commands** Electronics (such as transistor radio, televisions)

Manufacturing, such as in recording machines and voice recognition analyzer.

b) Operations Services (such as in Hospitals, households and schools)

3) **Image Understanding.** (ROBOTS AND COMPUTERS)

a) From Images to Objects Models (Such as in schools, engineering, farming, mining, hospitals and business).

b) Computing Edge distance recognition (Such as in engineering.

c) Interpretations of Images and surface Direction. (Such as in farming, engineering, Hospitals and business).

Mathmatetical Expert Systems Analysis and Education

ARTIFICIAL INTELLIGENCE IS CONCEIVED THROUGH

THE FOLLOWING MODES OF LEARNING:

1). Learning class descriptions from

Samples.

2). Learning rules from Experience.

3). Learning form from functional

Definition

MATHEMATICAL MODULES ARE ALSO USE TO

FORMULATE THE FUNCTIONS OF COMPUTERS AND

ARTIFICIAL INTELLIGENTS.

Mathmatetical Expert Systems Analysis and Education

EDUCATION AND SCIENTIFIC METHODS

Mathmatetical Expert Systems Analysis and Education

Education is through motivation and satisfying the needs of humans. The scientific world is part of an elongated human development. This this can be substantiated with the use and evolution of technologies. Education of the entities that comprise the need to achieve the ultimate goal of science is an important issue of today. Education is a conglomerate of beliefs of individual mind.

The scientific world plays a role in the development of education. It can be said that science is based on strategic planning. In the early evolutionary stage, it has been noted through history that ideas and inventions can be obtained through exploration and scientific abomination. Education is the foundation of the continuity sustainability and transformation. The group of individual learners can be the sole of success of education of mankind.

We can achieve our needs through critical innovation of the mind regardless of our roles in society. Everyone is a learner since we do not have control over what is to be learned. The circumstances surrounding education and its mode of delivery may be due to affordability and

Mathmatetical Expert Systems Analysis and Education

security. These in turns affect the volatility and the flexibility of learning.

To eliminate doubts and worry, education needs to justify the prosperity of societal factors of individuals. The incumbents involve must have the resources of attaining their goals. Since we have various goals and needs, the society or organizations must always embed or include scenarios and standard of accomplishments with their expectations.

The modalities of learning comprised all entities of understanding processes of humans. The dexterity of the mind can be explained through all means of communications. Both internal and external modes of communication can be justified in the development of intelligence.

The learning processes not only consist of spiritual processes, but all physical environmental and technological scientific means. The learning modes changes as one progresses through the channel of dwelling of living. The society must have realized that learning yield success only if it is applied substantially through the minds of the individual beings. Individuals learned mostly under the assumptions that they

Mathmatetical Expert Systems Analysis and Education

possess already all the preliminary process of life within the society. As mentioned in one of the seminar sessions in Tucson, Arizona, we as humans tend to follow with a can-do attitude. Acquiring self-knowledge always demand self-reflection. There is no way we can get to know ourselves if we don't take some quiet time to meditate. Contemplation is another one of the ways people tends to learn. Most people are willing to open to ideas and will try untested approaches and accept risk of learning. When people are at their personal best, their projects involve creative thinking and beyond-the-boundaries thinking because of the atmospheric conditions accord to them during the process of learning.

Even though we have gone through a process of learning at an early stage, we must have realized that nothing is done perfectly the very first time, not in schools, not in sports, not in games and certainly not in communities. We must also understand that we as humans evolve through changes, humans tend to search for learning opportunities. Opportunities that will meet the current changes and the foresee changes. The

Mathmatetical Expert Systems Analysis and Education

future of science depends on the learning materials of the present.

Changes may involve physical, psychological and social changes as opposed to environmental changes in our society and schools. Education and technology rest on the hands of the beholder. Education and science must be intrigue in our mind as important aspects of life, as we progress through life. Education and science are based on needs and consequences derived from the pasts. We all make mistakes and we must learn from our mistakes which is a form of making progress.

Education and Science is based on the homogeneity of physical resources available to us as human. Our adaptation is the objectivity of our consciousness. It should be noted that contemptuous circumstances can be resolved through education.

Educating the mind is prolific; we should engage in the learning process in all avenues. Education is a process whereby we should all learn together regardless of whom you are or where you come from.

Community, identity, stability is the main characteristic of the education methodology in the society. As it is often conveyed in parabolic ways, stability is required of

Mathmatetical Expert Systems Analysis and Education

any individual if he or she is to succeed in the society. And in order for individual to portray a positive identity within the society educational stability will have to play a vital role in acquiring knowledge. As it is often said, Knowledge comes from learning and experience while learning and experience are respectively derived from trying and doing. Without stability and knowledge, it may be impossible to acquire success. Individual may quest for knowledge, stability and success at early stage of their educational career, but these entities may later be suppressed at a later stage of their life.

The possibility of attaining all the individual's goals may rest solely on the individual and the society at large. The learning process depends on the motivational level of the individual which may encompass the ingredients of success. The ingredient of success in the society may determine the notions of knowledge and experience. Education as an art to prowess is the basis of for integrity in the society as whole. The power of success is achievable through knowledge. The initial educational attributes of individual suffice as learning takes place. Knowledge based on experience at an initial stage may result in learning activity of the present.

Mathmatetical Expert Systems Analysis and Education

Learning enterprise is a mode by which individual survived in the society. Accumulation of wealth may have to depend on knowledge and experience in the society; learning and success inevitably juxtaposed the amenities of wealth. The essence of educational training is the preparedness of individual to stability and success. It must be addressed to the problematic situations of individual in the society.

The circumstances surrounding propagation of learning is not solely materialism, but on the gratitude of knowledge. The standard which knowledge and materialism is attained is the repertoire of the educational establishments. In rationalizing the commonwealth of training individual, the society should apply transformation and sustainability in the evolution of education and science. The extenuation of objectives may depend on current and past activities. The educational solitudes may result in self-actualization of goals and thereby create self-awareness.

The technicality of learning may be justified by the scope of activities in the society. Education of the literate is different from that of illiterates in the society. Literacy does not mean everything is known, there are lessons to be learn from everyday activities in the society. Illiteracy of the mind is tolerable in certain aspect of learning.

The integration of learning may depend on the theme that individual needs, to know the anthology of survival. Stability projects the purpose of learning new ideas in our world. The determination of success rests on stability and knowledge.

Education of the mind is congenial to the cognitive approach of learning in the environment. It is believed that constant attention to the mind may gear up the learning processes. Educating the mind is a process whereby all activities are concentrated on the purposes of achieving positive results. Everyone must yield to the proliferation of the audacity to learn new ideas to attain success in the society.

There is a correlation between what is technology of the past and the present. The technological planning is based on what constitute technology in the modern society Today Information age products are based on knowledge, values and pragmatic approach. The educational approach is what constitutes technological management tools. Products management is of great

Mathmatetical Expert Systems Analysis and Education

importance as well as information management in

globalization. The past and present terms are reciprocity

of the global markets.

The retention of products serves the purpose of

conservatism and pragmatic values and approach.

Change of products and information are preserved in

our institutions. This change is pragmatic to the

management of the institutions or societies. The

information age is a co-existence of values and desire.

The philosophical entities composed of knowledge

resources. The composition of information is the entity

of products in the global market. The consistency of

products is based on the magnitudes of the information

obtained from the past and present.

Models and consistencies are aligned with

extensive extenuation of objectivity in technology. With

the extenuation of objectives, there are subjectivities to

innovations. The past may be subjective to the

development of the modern. Products and information

are inter-changeable commodities in technological

society. The aspect of globalization is empowered

through local development of external entities. The

Mathmatetical Expert Systems Analysis and Education

adaptability of external local entity serves as

philosophical global entity

The configuration of external foreign entities,

such as tools of the past era results in commonwealth

of technology. Technological posterity serves as the

philosophical view of the modern. As a result of

technological innovation there is a prowess in the dark

of perspective and convenience. Information are

internally and externally exposed and stored

technologically. The value of information is the logical

dwelling of philosophical abstracts of technology.

The transcendence of technology depends on

innovation of both present and past eras in order to

attain sophistication and adaptability. This presents the

future with lasting and endurance of educational tools in

our institutions. Technology is adeptly the invigoration

of educational expertise in our society. The

combinations of know how, when and why will be

attributed to the development and enhancement of

technology tools and their awareness.

We must adhere to the improvements of the

previous and the modern, as well as gear towards new

developments in our institution. Education serves as a

Mathmatetical Expert Systems Analysis and Education

purpose for improvements of knowledge and the

physical perspectives of the well-being of individual.

The institutions of higher learning undergo changes in

the light of technological innovations and the call for in

depth knowledge of the circumstances. Thus education

is a continuous and infinitely eloquent subject in our

institution and in our society as a whole.

Mathmatetical Expert Systems Analysis and Education

RELATIONSHIP BLOCK REPRESENTATATION

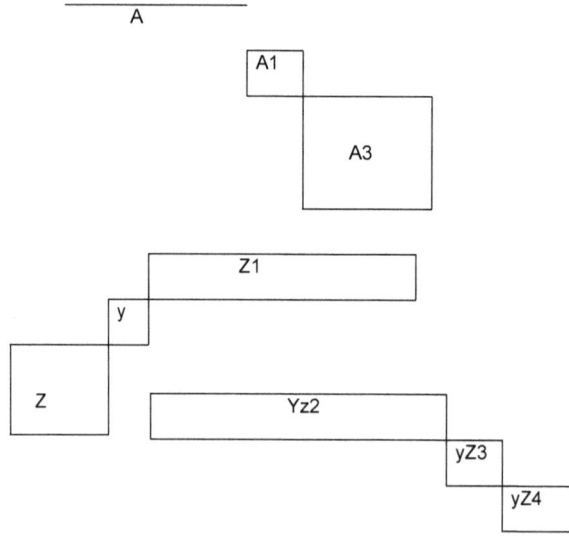

Mathmatetical Expert Systems Analysis and Education

$$A = (a1 + y1 + y3 + z4)$$

$$Z = (a1 + y1 + y3 + z1 + z2 + z4)$$

$$Y = A + Z$$

REPRESENTATION OF DATA

Elements(x)	1	2	3	4	Total
	5	7	8	16	36 (Z)
View (y)	2	6	8	16	32 ()
	7	8	8	16	36 (z

	A					
	B		C	D	E	F
	Total(f)					
	32	3	1	0	0	

MATHEMATICAL ANALYSIS

The above chart indicates that there are four basic elements which consist of 36 different objects of which 32 are very good, 3 good, 1 fair, 0 excluded.

$$x1 + x2 + x3 + x4 = Z$$

Mathmatetical Expert Systems Analysis and Education

$$y1 + y2 + y3 + y4 = z$$

$$x1 - y2 = C$$

$$x2 - y2 = D$$

$$x3 - y3 = E$$

$$x4 - y4 = F$$

$$A + B = t$$
$$Y0 = Y1 + Y2 + Y3 + Y4 = t$$

ROTATIONAL DELAY

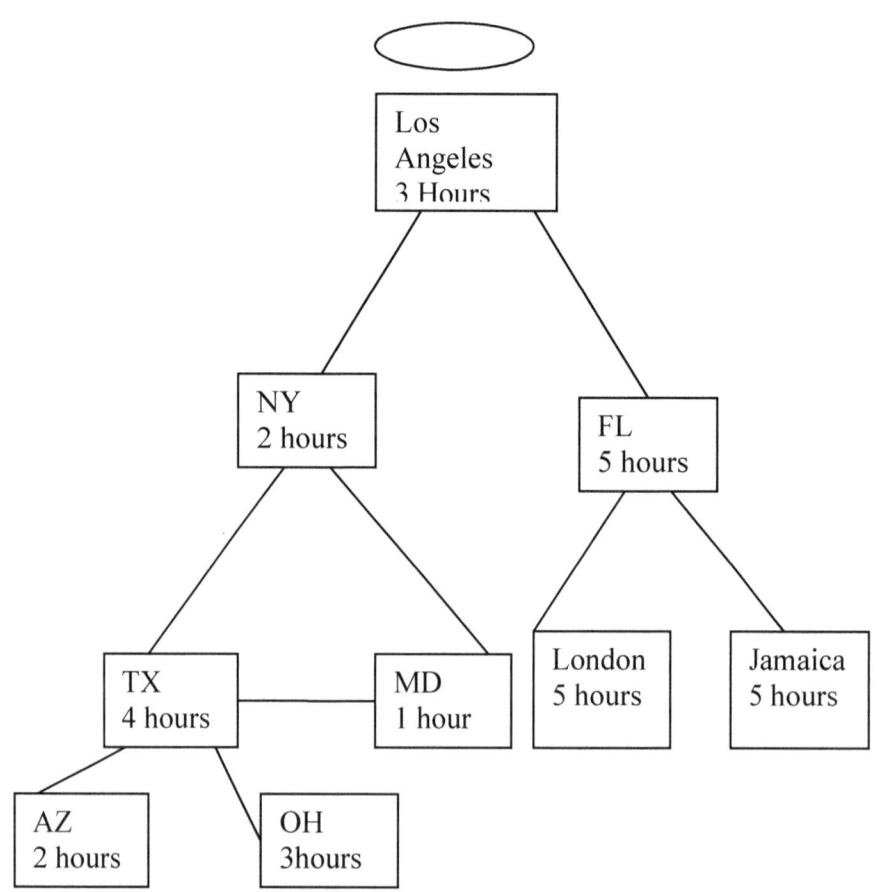

THE STRUCTURAL VIEW

Observations and Inferences in words associated with
Spaceship or Plane:

Luggage   Host

Pilot   Compass

Crew   Clients/Astronauts

Space   Cloud

Sky   Air

Mathmatetical Expert Systems Analysis and Education

INFERENCES AND OBSERVATIONS

Please observe and make inferences in words associated
with Spaceship or Plane:

Luggage ----------------------------------
-

Pilot ----------------------------------
--

Crew ----------------------------------

Space ----------------------------------

Sky ----------------------------------

Stars ----------------------------------

CLASS--------------------

Date.................

Name...
....

Instructional Development vs. Instructional Delivery

There is actually a difference between developing instructional materials for learners and delivering instructions for learners. The development of instructions may be well focused, yet the method of delivery is out focused or not well established as its form of presentations. As one develops the instruction, it is very important to stay focused on solving the performance problem when it comes to delivering the instructions. We must make sure that the objectives support the resolution of the instructional needs. One should always take into considerations either the content or skills that will help the learner improve and the mode of delivering the instructions should also be emphasized in the instructional development.

The Instructional Lecturing and Discussion Strategies

One of the most effective ways of delivering instructions is the combination of lecture and discussion method. The lecturer should always remember that learning is enhanced when learners are actively involved. In other

Mathmatetical Expert Systems Analysis and Education

word, participation is the key to learning process. This makes very important to get the learners involved during lectures. As the saying goes, "Get involved in it to get something from it." This means that it is therefore very important to develop a plan for including learner participation activities when lecturing. In facilitating learners' understanding of the material, lectures and discussion should also be clear and well organized.

Lectures as viewed by the majority is physically delivering of instructions or speaking to someone or to a group of individuals in a designated place, be it in a school, church or cooperate establishment. The set of instructions or statements should not be presented solely by actually speaking to audience; they should always involve both discussions and visual illustrations of topics to be lectured.

The initial presentation for an objective may provide the learner with information needed to achieve the main objective. For example, the initial presentation of a concept will include the concept name, definition and best examples – the lecturer task analysis therefore should in fact include this information.

The traditional mode of instructional delivery is by lecture. Although various modes such as distance education,

Mathmatetical Expert Systems Analysis and Education

in which video monitors, TV cameras or microphones, along

with other optional equipment (e.g. a fax machine) is being

used as the two-way transmission of voice have since

evolved from traditional method of lecturing. As discussed

earlier in the text, lecturing and discussion methods are

instructional methods or strategies that can be used most

effectively in the delivery of instructions. In lecturing, the

instruction actually tells, shows dramatizes, demonstrates or

otherwise disseminates subject content to a group of

learners. This pattern of instructional method can be utilized

in a classroom, an auditorium or a variety of locations

through the use radio, amplified telephone, close circuit

television transmission, interactive-distance television or

satellite communication (Teleconferencing).

Strengths of Lecturing and Discussion

Since discussion offers the opportunity for a good deal

of learners' activity and feedback, it should, according to

theory, be more effective than lecture in developing

concepts and problem-solving skills. The benefits of using

lecturing and discussion method to accomplish certain

learning objectives can be enumerated as follows:

- Lecture and discussion format is very familiar and
  conventionally acceptable to instructors and learners.

Mathmatetical Expert Systems Analysis and Education

The lecture method is, according to Morrison, Ross & Kemp (2001), the most common form of instructional delivery.

- Lectures can be easily quickly designed since the instructor is familiar with the material and will make the actual presentation. The designer of lectures usually provides the instructor with a list of objectives and a topic outline. The assumptions usually in mind is that the instructor can make necessary strategic decisions.

- Lectures and Discussion place the instructor in direct control of the class and in visible authority position. For some instructors and in many teaching contexts, the factors are advantageous for achieving the objectives.

- With a lecture, large number of learners can be served at one time. The group of learners is only limited by the size of the room; therefore, lectures can be highly economical.

- Lectures and discussions can be feasible methods of communicating when the information requires frequent changes and updates or when the information is relevant for only a short time period,

such as the implementation of a new travel policy.

- It should be noted that a good lecture and discussion

  can be motivating or interesting for the learners.

The effective use of Lecturing and Discussion

Discussions can be effectively used among matured

learners and with a small group size of learners whereas

lecturing is the opposite. Lecturing is most appropriate when

there is a large group size of learners. Lecturing in actual

sense should be augmented with discussions and

illustrations.

The feelings of the learners should be taken into

considerations when preparing and delivering lectures.

Feeling actually included is the key to learner motivation to

learn. It is very important then that learners not encounter

overt hostility, ignorance, and insensitivity, as well as subtler

messages that their cultural heritage is not valued. The

message received always is that there is an underlying

resentment about the pre-sentence of the learners who

"don't fit." Many well-meaning but ignorant statements are

the cause for discomfort in lecturing. In lecturing, we must be

Mathmatetical Expert Systems Analysis and Education

patient with ourselves and the learners while several

generations try to unlearn some deep-seated prejudices.

While lecturing, all learners need to feel welcome.

They need to feel that they are being treated as individuals.

They also need to feel that they can participate fully. The

need to be treated fully well is another message that needs

to be addressed when delivering lectures.

One thing lecture is good for is that it can provide

structures to help learners read more effectively. In facts

lecture may help learners learn to read quickly. Readability

of material depends on expectations brought to the material

by the reader; therefore, better or innovative lectures can

build structures and expectations that help learners read

material in the given subject matter area more effectively.

Lectures can also have indirect values apart from their

cognitive content. Many lecturers today have motivational

functions. By helping learners become aware of a problem

as of challenges to ideas they have previously taken for

granted, the lecturer can stimulate interest in future learning

in an area. Moreover, the lecturer's own attitudes and

enthusiasm have an important effect on learners' motivation.

Not only is the lecturer a model in terms of motivation and

curiosity, the lecturer also models ways of approaching

Mathmatetical Expert Systems Analysis and Education

problems, portraying a scholar in action in ways that are

difficult for media or methods of instruction to achieve.

Finally, there are values in lecturing for professors

themselves.

Mathmatetical Expert Systems Analysis and Education

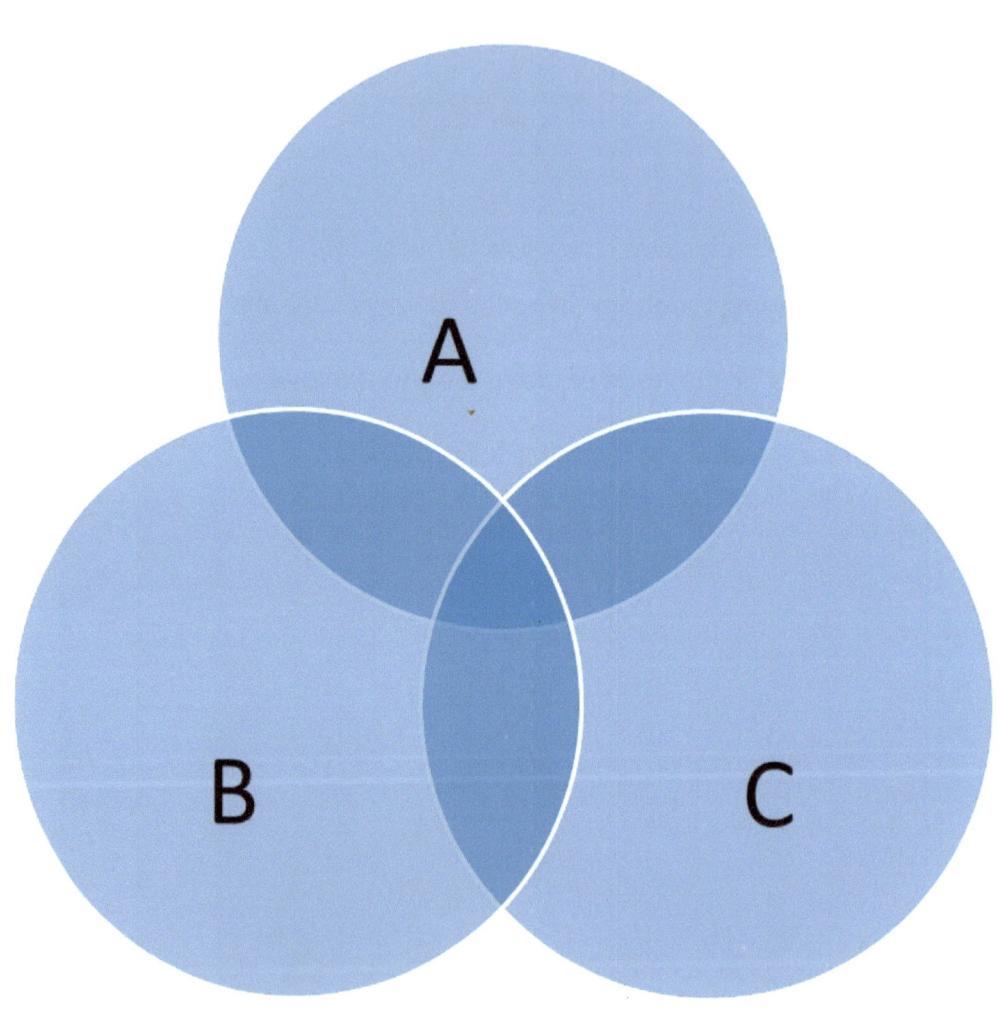

A   ↔   LEARNERS

B   ↔   DICUSSIONS

C   ↔   LECTURES

MOTIVATIONAL CONDIMENTS OF LEARNING

Learning enterprise is a mode of which individual survived in the society. Accumulation of wealth may have to depend on the knowledge and experience in the society, learning and success extend beyond the amenities of wealth. The essence of educational training is preparedness of individual to stability and success. It must be addressed to the problematic situations of individual in the society.

The circumstances surrounding propagation of learning is not solely materialism, but on the gratitude of knowledge. The standard which knowledge and materialism is attained is repertoire of educational establishments. In rationalizing the commonwealth of training individual have to be instituted in transformation and sustainability of the evolution of education. The extenuation of objectives may depend on current and past activities. The educational solitudes may result in self-actualization of goals and thereby create self-awareness of the episodes.

The technicality of learning may be justified by the scope of activities in the society. Education of the literates may be different from that of illiterates in the society. Literacy does not mean everything is known, there

Mathmatetical Expert Systems Analysis and Education

are lessons to be learn from everyday activities in the society. Illiteracy of the mind is tolerable in certain aspect of learning.

The integration of learning may depend on the theme that individual need to know the anthology of surviving. Stability projects the purpose of learning new ideas in our world. The determination of success rests on stability and knowledge.

Education of the mind depends on the cognitive approach of learning environment and individual. It is believed that constant attention to the mind may gear up the learning process. Educating the mind is a process whereby all activities are concentrated on the purpose of achieving results. All learners may have to yield to the proliferation of learning new ideas to attain success.

THE KOLB LEARNING THEORY

The Kolb' theory of learning styles, which can be termed as "experiential learning" is actual based on the model of social learning. In fact, Kolb's theory came about as a result of the works performed by Dewey, Lewin, and Piaget. The phases of Kolb's theory can be enumerated as follows:

Mathmatetical Expert Systems Analysis and Education

The learners become immensely involved in a solid experience of their own.

The learners learn by reflections and observations of self and others.

Abstract concepts and generalizations are symmetrically formed by the learners through inductive reasoning and not deductive reasoning.

The concepts and generalizations that have been formed by the learners are empirically tested and experimented on by the learners which eventually lead to deep involvement in solid experiences.

According to Kolb, there are four different types of individuals in the experiential learning process. These individuals can be classified as:

Diverges: These categories of individuals actually transform the experience through reflective observation by taking in the experience through solid experiences. The "diverges" view all the different aspects of a situation and then combine them into a meaningful one.

Assimilators: Abstract conceptualization is injected by these categories of individuals and it is transformed through reflective observations. The assimilators are good in creating theories and converting the diverse data into integrated whole.

Converges: These Categories of learners take in the experience through abstract conceptualization and transform it through active experimentation. According to Kolb, these individuals are able to consume ideas and come up with the correct solutions.

Accommodators: These individual learners consume the experience through solid experience and transform it through solid active experimentation or participation.

The accommodators are actually risk takers, according to Kolb, who are able to adjust to new situations and this type of learners use trial and errors methods in learning subjects or topics.

The Importance of Kolb's Contributions to Improving the Learning Process

Kolb's contributions in recognizing and providing for varying learning styles and improving the learning/teaching process are of great importance to education. Kolb's contributions actually justified the psychological and the physiological processes of learners acquiring knowledge about their ability to think and learn from experiences. The

Mathmatetical Expert Systems Analysis and Education

four phases of Kolb's learner modes are very important to

the learners' success of acquiring knowledge. Observations

of self and others play a vital role in learning and teaching. It

can be further stressed that through observations of self and

others, learners are able to assimilate and converge what

they learned.

Reflection is a way to convey to the individual learners

what have been learnt, and it also enables the learners to

make assumptions as to what is to be learnt. The Kolb's

theory is a cognitive style of learning. It is

also a preferred physiological modality of learning and

teaching. With the Kolb's theory, the strategic learners need

to be able to set and use meaningful goals to help them

generate and maintain their motivations for learning.

Providing learners with appropriate opportunities to reflect

on the subjects or topics under discussion might be useful to

them as they strive to reach their personal, social and

occupational goals. Skills are very important in the learning

process. Experiences and skills are both the conjectures of

knowledge. Learners must want to learn if they are to

acquire the knowledge of the subjects. "Skill and Will"

actually results in self-regulated learning. "The longevity of

the learners is the homogeneity of his body, and the

Mathmatetical Expert Systems Analysis and Education

spontaneity of his adaptability is the objectivity of his

consciousness".

KOLB LEARNING/TEACHING STYLE

Example of Kolb's Theory in Educational Setting

An example of Kolb's theory of learning in an educational

setting is where the learners engage in a laboratory work. A

laboratory is set up where knowledge is put into practice.

The processes of assignments or works are observed and

reflected upon in a practical manner. Another example is the

internship program adopted by various schools, universities

and colleges. Learners are able to bring to the classrooms,

experiences from the work place and to the work place,

experiences from the classrooms. These methods of

learning evenly provide cognitive learning styles. Testing is a

way we can also relate to Kolb's learning theory. It is a way

to measure what has been learnt by experience. Experience

may be through studies or participating in actual activity. In

an actual sense,

testing is an experience because the learners get to express

themselves and diverge and assimilate what has been

learnt.

Mathmatetical Expert Systems Analysis and Education

Representation of Kolb' s Learning/Teaching Theory

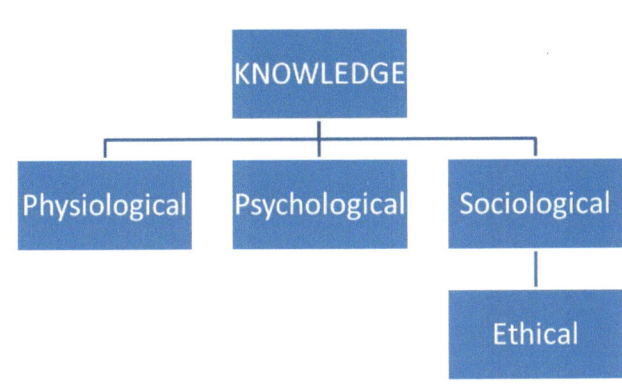

Fig. 1:   The Learning/Teaching   Modalities

Application of Kolb Theory to Computer Instructions

The Kolb's theory of learning can be practically and
adeptly applied to the computer instruction. Learners of the
subjects, i.e. computers, are obliged to actual experience of
the practical notions of the subjects. Concepts and
generalization of the topics must be made visible and
adaptable so the learners can reflect, diverge, assimilate and
accommodate the learned process. The learner analysis and
contextual analysis should be viewed easily as identifying
constraints to learning and teaching. Experience of the
subjects can be attained through repetitive practice of the
learned processes. Divergence and convergence of the

Mathmatetical Expert Systems Analysis and Education

concepts are mandatory consequences of learning the

subject.

===
===
===
===
===
===
===
===
===
===
===
===
===
===

www.ingramcontent.com/pod-product-compliance
Lightning Source LLC
Chambersburg PA
CBHW040826180526
45159CB00001B/79